647

MAKE it WORK!

INSECTS

Wendy Baker & Andrew Haslam

written by
Liz Wyse

Photography: Jon Barnes
Series Consultant: John Chaldecott
Science Consultant: Duncan Reavey
Lecturer in Biology at the
University of York

Scholastic Canada Ltd.,
123 Newkirk Road, Richmond Hill, Ontario, Canada L4C 3G5

MAKE it WORK!
Other titles

Earth
Electricity
Machines
Plants
Sound

Published in Canada in 1994 by
Scholastic Canada Ltd.,
123 Newkirk Road, Richmond Hill, Ontario, Canada L4C 3G5

First published in Great Britain in 1993 by
Two-Can Publishing Ltd
346 Old Street, London EC1V 9NQ

Printed and bound in Hong Kong

2 4 6 8 10 9 7 5 3 1

Canadian Cataloguing in Publication Data

Baker, Wendy
    Insects

(Make it work!)
ISBN 0-590-24204-0

1. Insects - Juvenile lierature. 2. Insects -
Experiments - Juvenile lieterature. I. Haslam,
Andrew. II. Wyse, Liz. III. Barnes, Jon.
IV. Title. V. Series: Baker, Wendy. Make-it-work!

QL467.2.B35   1994   j595.7   C93-095147-6

Editor: Mike Hirst
Series concept and design: Andrew Haslam and Wendy Baker
Additional design: Helen McDonagh
Thanks also to: Rachel, Catherine, Holly, Marco and Lee;
Colin and Jenny at Plough Studios.

The photograph on page 11 was taken by Stephen Dalton (NHPA).

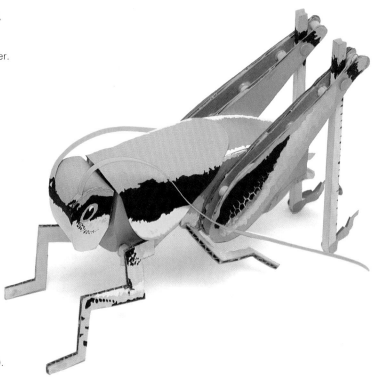

Contents

Being a Scientist   4

Parts of an Insect   6

Looking at Insects   8

Insect Sight   10

Insect Antennae   12

Jumping Insects   14

Insects in Flight   16

Dragonflies   18

Insects Feeding   20

Insect Defence   22

Insect Enemies   24

Beetles   26

Water Bugs   28

Butterflies   30

Metamorphosis   32

Moths in the Night   34

Flexible Insects   36

Bees   38

Ant Colony   42

Insect Architecture   44

Glossary   46

Index   48

Words marked in **bold** in the
text are explained in the glossary.

Did you know that there are more kinds of insects on the earth than all the other animal **species** put together? Insects are small, and most of them don't live for long, but they are extremely tough. They can be found in almost every type of **habitat**, they breed in huge numbers and have clever ways of tricking and avoiding their enemies.

Entomologists are scientists with a special interest in insects. They observe the insect world and try to find out how it works. They investigate where insects live, what the different species look like, and how insects behave.

MAKE it WORK!

There are insects all around us: in our homes, in gardens and out in the countryside. You don't need a science lab to become an entomologist. Try the activities in this book and you'll soon be an insect specialist yourself.

Seeing insects

Many insects are very small, and you will need help to look at them closely. In this book, the photographs of some insects are magnified so you can see them better. The number at the side of the name label tells you how much bigger than real life the photograph is.

greenbottle

giant long horn beetle

Bug boxes

You can use a bug box to help you look at tiny insects. It's a small container with a magnifying glass in the lid. If you don't have a bug box, you could simply put an insect inside a matchbox and examine it through a magnifying glass.

Catching insects

You may want to catch insects to look at them more closely. A butterfly net is the best way of catching flying insects, but make sure it has a fine mesh, so it doesn't hurt the trapped insect. A **pooter** is another clever insect-catching device, which lets you suck insects into a jar.

Remember!

Always follow the insect spotter's code (on page 8) and always let insects go again after you have looked at them.

butterfly

earwig (x2)

Making observations

Here are some questions to ask yourself when you spot an insect:
- What does it look like?
- Has it any unusual markings?
- What is it doing – feeding, resting or fighting?
- What sort of habitat is it in?
- What is it eating?

rhinoceros beetle

Recording your findings

To be a good scientist, you should make a record of everything you've seen and done.

Field notebook When you are out and about looking at insects, you should always carry a field notebook. Use it for writing down all of your observations.

Camera Photographs will give you an exact record of what an insect looks like. But be careful when stalking insects with a camera – they are easily frightened!

Small tape recorder You can use a small tape recorder for taking notes too. Speak into it as you look at the insects, or try to record the sounds made by the insects themselves.

Entomologists divide insects into different kinds or species. Every species also belongs to a bigger group, called a **genus**. *All insects have Latin names. The first part is an insect's genus, and the second part is its species. The common black garden ant is called* Lasius niger *in Latin.* Lasius *is the genus, and* niger *shows which species of* Lasius *it is.*

Sumatran beetle

6 Parts of an Insect

How can you tell if a creature is an insect? It's easy! All insects have the same main body parts in common.

Insects' bodies are divided into three parts: the head, **thorax** and **abdomen**. They have three pairs of legs, all attached to the thorax, and on every insect's head are a pair of **antennae** and two **compound eyes**.

MAKE it WORK!

Make a model insect, and see for yourself how the parts of its body fit together.

You will need

pipe cleaners	wire
a piece of polystyrene	tracing paper
a wooden kebab skewer	10 wooden beads
acrylic paint, or poster paint mixed with a little rubber-solution glue	

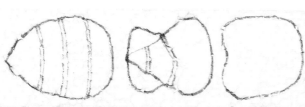

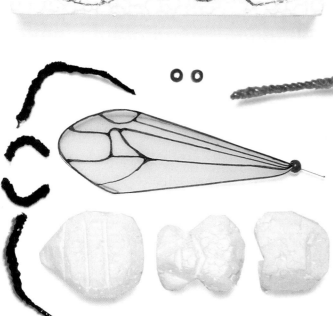

Be careful! Some of the activities in this book use polystyrene shapes. It's best to cut the polystyrene with a special electric polystyrene cutter. If you don't have one, ask an adult to help you use a craft knife, and take great care with the sharp blade.

1 Draw the head, thorax and abdomen shapes on the polystyrene, and cut them out.

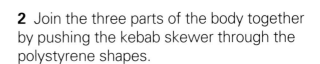

head

antennae

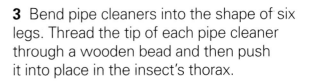

2 Join the three parts of the body together by pushing the kebab skewer through the polystyrene shapes.

3 Bend pipe cleaners into the shape of six legs. Thread the tip of each pipe cleaner through a wooden bead and then push it into place in the insect's thorax.

It's easy to think that all creepy crawlies are insects, but real insects have a special body structure that makes them unique.

*The creatures on the right are **not** insects. The spider has only two body parts, and it has eight (not six) legs. Centipedes and millipedes have many body segments, with legs on each segment!*

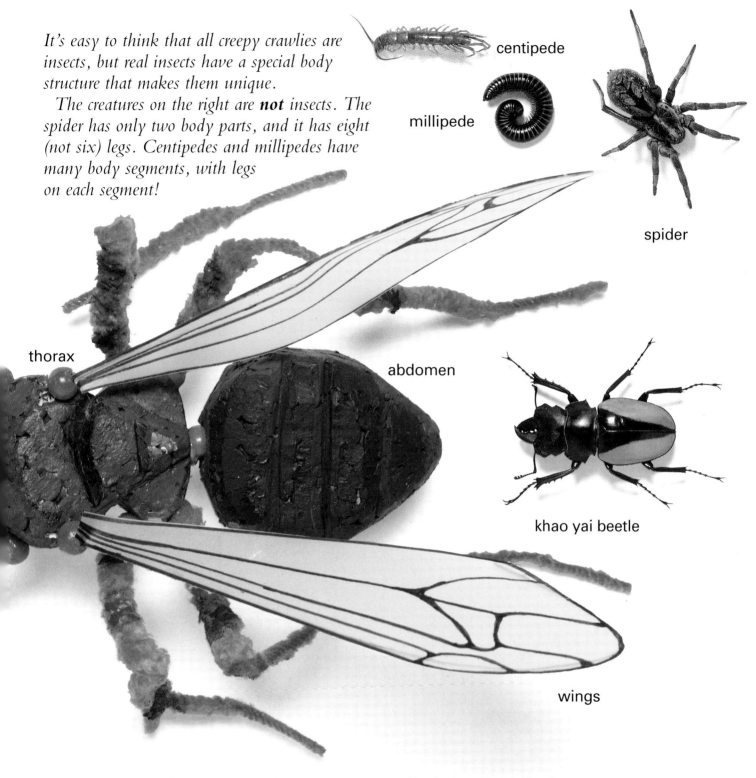

centipede

millipede

spider

thorax

abdomen

khao yai beetle

wings

4 Paint the parts of the body. Acrylic paint is best, but you could also use poster paint mixed with a little rubber-solution glue.

5 Cut the tracing paper into wing shapes as shown and paint a vein pattern on each wing. Bend the wires into wing shapes and glue them around the edges of the tracing paper.

6 Push the wire ends of each wing through a wooden bead and into the thorax. (Wings and legs are always joined to the thorax.)

7 Add pipe cleaners to the head to make the antennae and mouth parts. Finally, glue two wooden beads to the top of the head as the compound eyes.

8 Looking at Insects

You can learn a lot by looking closely at insects, but most insects won't help you by standing still! The best way to examine them is to catch them, look at them and then let them go again.

Always carry a magnifying glass, as most insects are small. Remember a notebook and pencil too, to jot down when and where you saw each insect.

The Insect Spotter's Code

● Always leave the environment exactly as you found it. For example, put back any logs and stones if you have been looking under them.

● If you are taking insects away, always make a careful note of what they are feeding on. Be sure you have enough supplies of food for your insect. If there is not enough food available, do not remove the insect.

● Always return the insects to the place where you found them.

MAKE it WORK!

Pitfall traps are a good way of catching insects that scurry along the ground. You can try out different kinds of bait to lure them in. You'll find that most insects like sweet things, such as rotten apples or a molasses and water mixture.

Small insects can be caught with a pooter. Suck them up into the jam jar and examine them carefully before you let them go again.

For the pitfall trap you will need

an empty yogurt pot   a piece of cardboard

1 Place the yogurt pot in a hole in the ground. Put your bait inside the pot.

2 Put four rocks around the pot. Place a piece of cardboard over the stones to add shade and keep the rain out. Weight it down with another stone.

stiff plastic tubing
a piece of gauze
a jam jar with lid
an elastic band
flexible plastic
 tubing
Plasticine
a bradawl

Making the pooter

1 Ask an adult to help you make two holes in the jam jar lid using a bradawl.

2 Cut two short pieces of tube, and one that is slightly longer. Fasten a piece of gauze over the end of one of the short tubes with a rubber band.

3 Push one short piece through each hole in the lid and seal the holes tight with Plasticine.

4 With the small piece of flexible tubing, connect the longer piece of tube and the other short tube. Screw the lid onto the jar.

5 Suck the short tube, pointing the end of the long tube at the insects you want to catch.

▶ What equipment should an insect spotter carry? With a butterfly net you can catch flying insects. A bug box helps you to examine them, and gloves will protect you against stings!

10 Insect Sight

Insects' eyes are very different from our own. They stick up out of the insect's head like the two big, round halves of a globe. Each eye is made up of many separate **lenses**, and every lens looks out on the world at a slightly different angle. The many lenses mean that insects can see a very wide area around them.

Insects can see both to the right and the left without moving their heads.

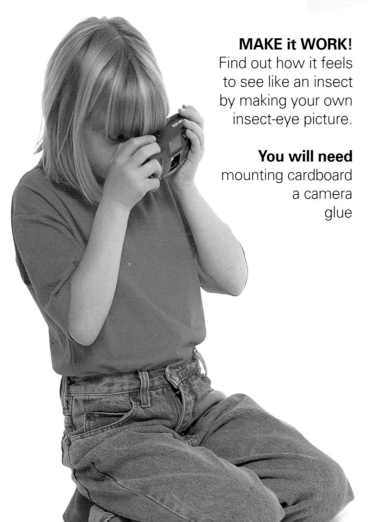

MAKE it WORK!
Find out how it feels to see like an insect by making your own insect-eye picture.

You will need
mounting cardboard
a camera
glue

1 Choose a subject for the picture. A big subject is best – a row of plants or a long building, for instance.

2 Position yourself in front of the subject. Check that the camera is in focus and then take the first photograph.

3 Turn your head a little, and take another photograph. Repeat this process several times, turning your head, but not moving your body.

4 When your photographs have been developed, mount them on a piece of cardboard to make one big insect-eye view of the subject.

MAKE it WORK!

How does it feel to know what's going on all around you without having to turn your head? Make a pair of insect eye glasses, and find out!

For the insect eye glasses you will need

balsa wood   strong glue
two small mirrors   a craft knife
thick cardboard (corrugated board is best)

1 Cut out the shape of the glasses in cardboard as shown. Fold along the dotted lines to make the arms, checking that they fit over your ears.

2 Cut three balsa wood triangles, each in exactly the same shape. The two shorter sides must be as long as the width of the mirrors.

3 Glue the mirrors to the balsa wood triangles, making a tent shape as shown. To make the shape stronger, add a cardboard base.

4 Glue the top edge of the mirror tent to the exact centre of the nose bridge of the glasses.

▼ When you put the glasses on, your field of vision will reach around to your ears – just like an insect's!

◄ The eyes of this horse-fly meet at the top of its head, so that it has almost all-round vision – very useful when hunting smaller insects for food.

12 Insect Antennae

Besides its eyes, every insect has two antennae – long, thin feelers that stick out from the top of the head. They are an insect's nose and ears, all in one! Near the mouth, some insects also have two smaller feelers, called palps, which taste food before it is eaten.

MAKE it WORK!

Make your own antennae. You can't smell or taste with them, but they do give you an idea of how insects are able to feel their way around.

You will need

a ball made of thin plastic
two thicknesses of wire
two table-tennis balls
coloured cardboard
glue and Velcro
two beads
sticky tape

blindfold

chin strap   headband

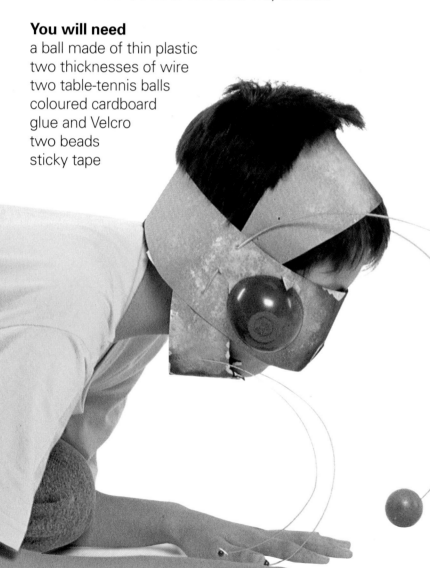

A crafty tip! For some of the activities in this book, you will need to cut out quite difficult cardboard shapes. If anyone has a photocopier you can use, you could copy the shapes that are shown in the book onto tracing paper, and enlarge them on the photocopier. Then cut out the photocopies to make full-size templates.

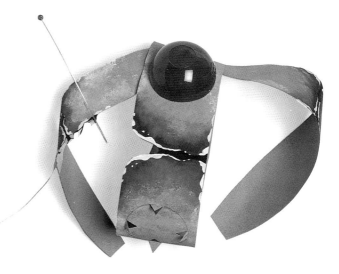

1 Cut out the three cardboard shapes shown in the photograph.

2 Ask an adult to help you cut the plastic ball in half. Glue the two halves onto the triangular folds on the blindfold, to make the insect eyes.

3 Stick Velcro pads on the blindfold. Try putting it over your eyes, and alter the position of the Velcro to fit your head.

4 Glue the chin strap and headband to the blindfold – make sure they'll fit your head too.

5 Take two pieces of thin wire and stick the beads to the top. Thread these into the chin strap as shown to make two palps.

6 Add the table-tennis balls to the tips of the thicker wire to make the antennae. With your mask in position, slide the antennae down each side of the blindfold. Fix them with sticky tape.

*Antennae are covered by tiny hairs. These hairs pick up sounds, smells and special chemical messages, called **pheromones**, that are sent out by other insects. Female moths produce chemicals which males can detect with their antennae from far away.*

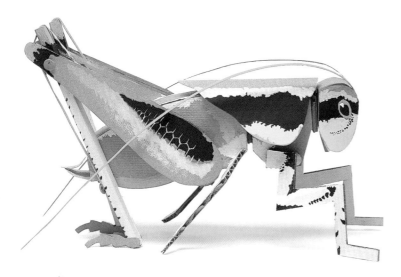

Although they are only small, insects are extremely strong. Jumping insects, such as grasshoppers, crickets and fleas, have long back legs with big, strong muscles to launch them into the air. The muscles are carefully protected because they are inside the insect's skeleton – different from our human muscles, which are on the outside of our bones.

MAKE it WORK!

The upper part of an insect's leg is the **femur**. Bush crickets have big femurs, to house their powerful muscles. Make a model bush cricket, with elastic band muscles, and watch it jump!

You will need

thin dowel
drawing pins
corrugated cardboard

elastic bands
coloured cardboard
small strips of wood

3 Join the three parts of the body by gluing small pieces of corrugated cardboard between them.

4 To make the femurs, strengthen the coloured cardboard with corrugated cardboard. Cut small pieces of dowel and glue them to the back of the femur shape as shown. Thread an elastic band around the two dowel pieces at the top of the femur.

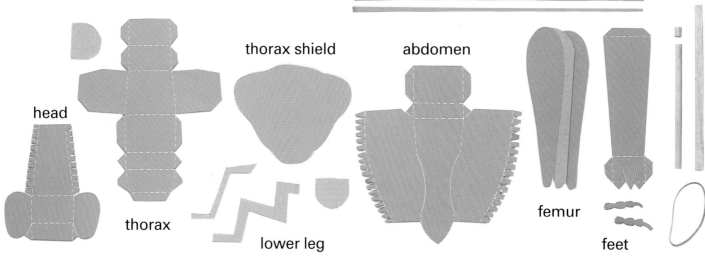

head

thorax

thorax shield

lower leg

abdomen

femur

feet

1 Cut out the cardboard in the shapes shown. Cut along the solid lines, fold the dotted ones and glue the tabs, to make box shapes for the head, abdomen and thorax.

2 The bush cricket's thorax is covered by a shield. Bend the shield over the thorax.

5 Glue another femur-shaped piece of cardboard to the dowel pieces to make the back of the femur.

6 Fasten the wooden rod to the femur with a thumbtack, as shown. This will make the lower part of the leg. Add another thumbtack at the back of the leg, near the foot.

7 Stretch the rubber band over this lower thumbtack. Hook it over the top of the leg rod.

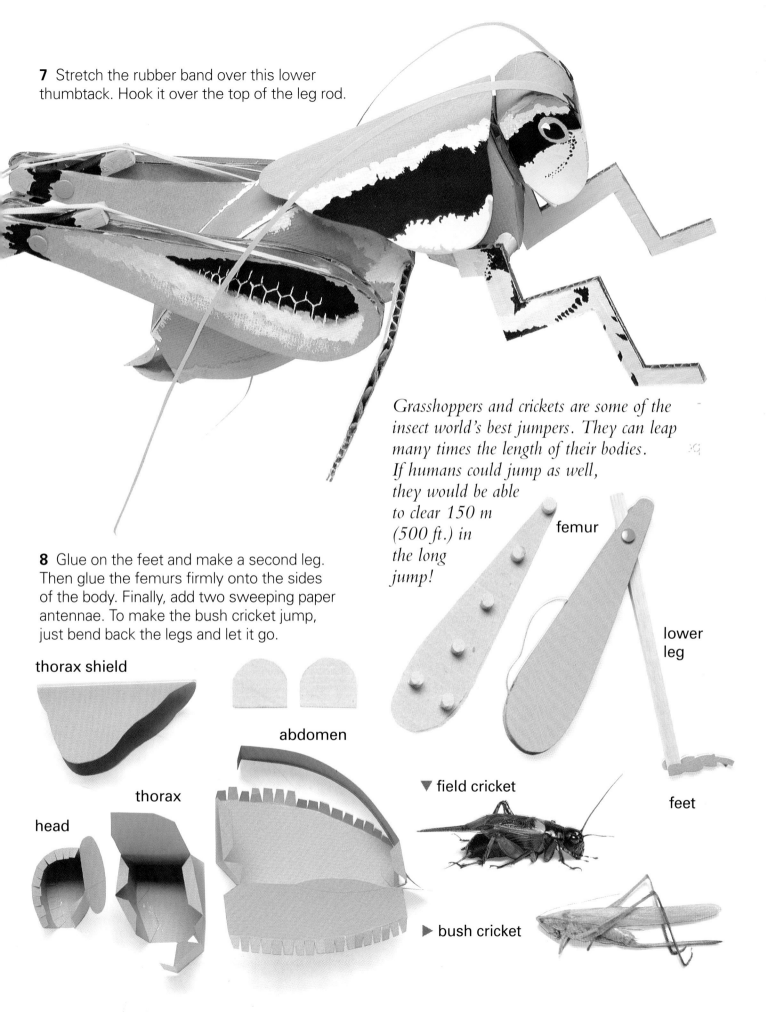

Grasshoppers and crickets are some of the insect world's best jumpers. They can leap many times the length of their bodies. If humans could jump as well, they would be able to clear 150 m (500 ft.) in the long jump!

femur

lower leg

8 Glue on the feet and make a second leg. Then glue the femurs firmly onto the sides of the body. Finally, add two sweeping paper antennae. To make the bush cricket jump, just bend back the legs and let it go.

thorax shield

abdomen

feet

▼ field cricket

thorax

head

▶ bush cricket

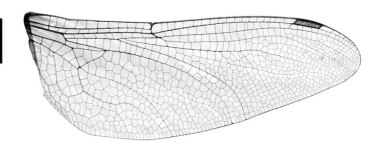

Many insects are able to fly, although they don't all move their wings in exactly the same way.

One of the most common groups of insects is actually called 'flies.' They have two wings which are moved by special flight muscles cleverly positioned inside the fly's thorax.

The surface of an insect's wing is covered by tiny veins that keep it stiff. Each species has a different pattern of veins on its wings. Scientists can use these different patterns to tell one species of insect from another.

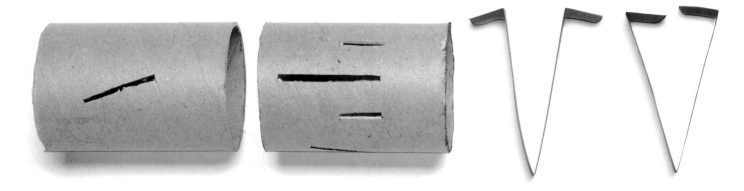

MAKE it WORK!

Make a model of a fly's wings and thorax. You will see how the two sets of flight muscles make the insect's wings move up and down.

You will need

thin red and blue cardboard
a craft knife and marker pen
plain cardboard, glue and sticky tape
a cardboard tube (eg. a kitchen-roll tube)

1 Using the craft knife, cut one slot at an angle on each side of the roll.

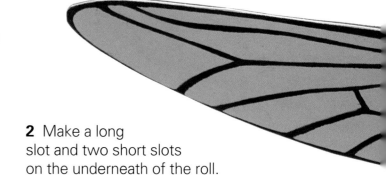

2 Make a long slot and two short slots on the underneath of the roll.

3 Cut two wing shapes as shown below. Draw the veins with a marker pen.

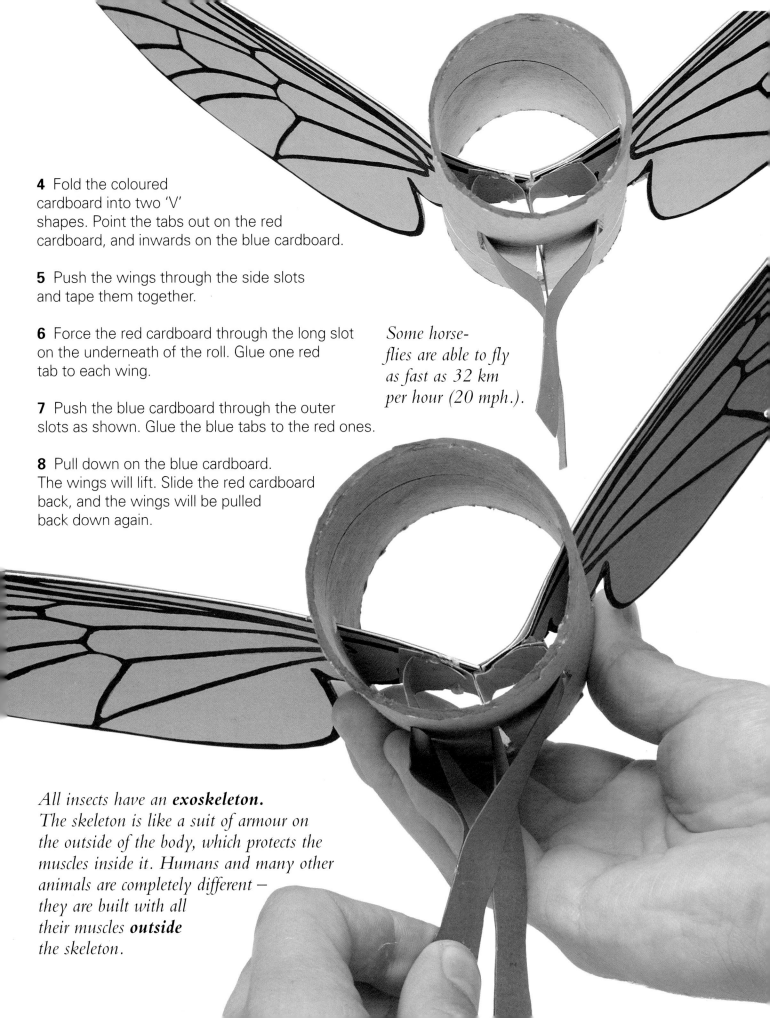

4 Fold the coloured cardboard into two 'V' shapes. Point the tabs out on the red cardboard, and inwards on the blue cardboard.

5 Push the wings through the side slots and tape them together.

6 Force the red cardboard through the long slot on the underneath of the roll. Glue one red tab to each wing.

7 Push the blue cardboard through the outer slots as shown. Glue the blue tabs to the red ones.

8 Pull down on the blue cardboard. The wings will lift. Slide the red cardboard back, and the wings will be pulled back down again.

Some horse-flies are able to fly as fast as 32 km per hour (20 mph.).

*All insects have an **exoskeleton.** The skeleton is like a suit of armour on the outside of the body, which protects the muscles inside it. Humans and many other animals are completely different – they are built with all their muscles **outside** the skeleton.*

18 Dragonflies

Dragonflies are the insect world's most skilled fliers. Their two pairs of wings move separately, like helicopter blades, so dragonflies can hover, fly backwards, and change direction at high speeds. They use their superb flying ability and powerful eyesight to catch smaller insects in mid-air.

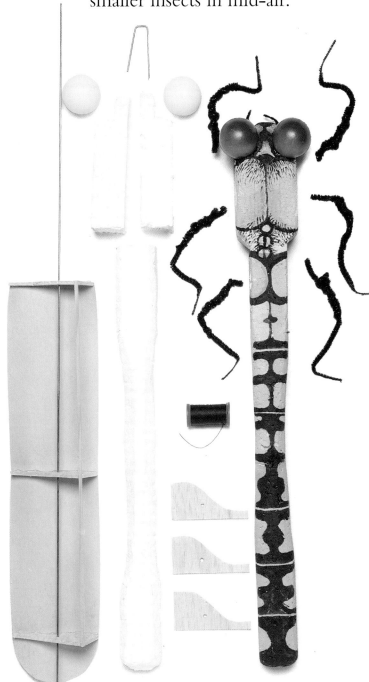

MAKE it WORK!

Make a model dragonfly. On a windy day you'll actually be able to fly it out of doors like a kite. You could also paint it with brightly-coloured markings, copied from a guide book to insects.

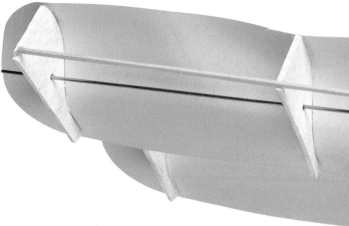

You will need

balsa wood	thin cardboard
pipe cleaners	strong twine
two table-tennis balls	sharp craft knife
a piece of polystyrene	strong wood glue

two thin metal rods, about 95 cm (36 in.) long
acrylic paint, or poster paint mixed with a little
  rubber-solution glue

1 Cut the body shape out of polystyrene as shown on the left. (See the safety note on page 6.)

2 Glue the table-tennis balls to the head to make the eyes, and paint the body.

3 Glue on six legs made out of pipe cleaners. Add two short pipe cleaners to the head to make the dragonfly's mouth parts.

5

7

4 Shape a short piece of wire into a hook. Push it into the underside of the body and glue it into position.

5 Ask an adult to help you cut six rectangles of balsa wood, about 10 cm (4 in.) wide. Cut each rectangle in two with a wavy line, to make the symmetrical shapes as shown.

11 Slide a wing onto either side of the rod. Run the rod through the holes in the balsa wood shapes so that the wings can spin freely.

12 Attach the other pair of wings in the same way. Bend each metal rod at the end so that the wings cannot slip off.

13 Tie a piece of twine to the hook, and you are ready to launch your dragonfly into the air like a kite. If it doesn't balance at first, weight down the tail with a little Plasticine.

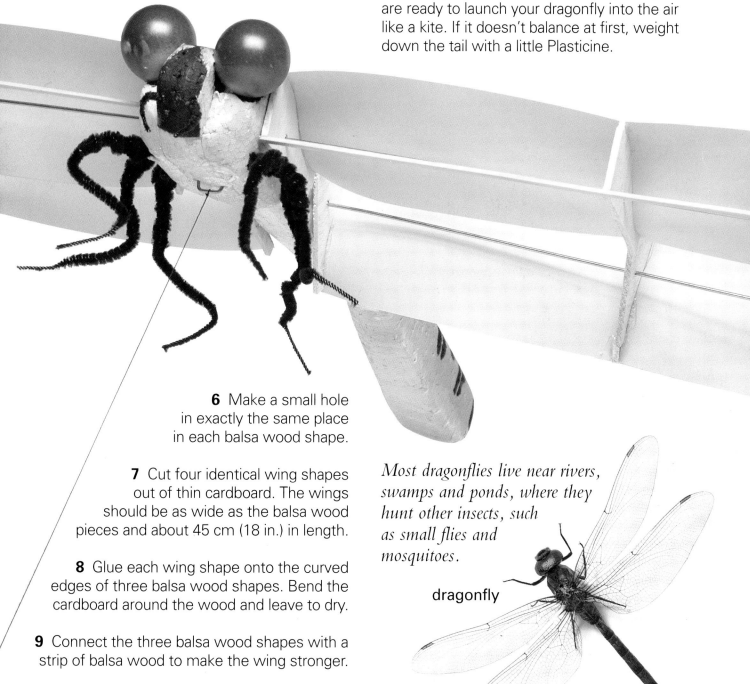

6 Make a small hole in exactly the same place in each balsa wood shape.

7 Cut four identical wing shapes out of thin cardboard. The wings should be as wide as the balsa wood pieces and about 45 cm (18 in.) in length.

8 Glue each wing shape onto the curved edges of three balsa wood shapes. Bend the cardboard around the wood and leave to dry.

9 Connect the three balsa wood shapes with a strip of balsa wood to make the wing stronger.

10 Ask an adult to help you push one of the thin metal rods through the dragonfly's body.

Most dragonflies live near rivers, swamps and ponds, where they hunt other insects, such as small flies and mosquitoes.

dragonfly

20 Insects Feeding

Most insects eat in one of three ways: by chewing, sucking or sponging. Their mouth parts are all specially adapted to the kind of food they eat.

Insects that chew their food eat like humans do. Their powerful jaws, or **mandibles,** crush, cut and grind the food. Insects with chewing mouth parts are the most common, and include dragonflies, grasshoppers and beetles.

It's important to keep food covered and away from flies. They spread diseases by sponging half-eaten food onto other food that humans may later eat.

fly (x 2)

MAKE it WORK!
Imagine what it is like to flick out a long, coiled proboscis to get your food, just like a moth or a butterfly. Make your own proboscis and see for yourself how it feels.

Sucking insects, such as moths or fleas, live off liquids. They feed on sap from plants or body fluids of animals. They pierce the food with a long **proboscis**, or feeding tube, and inject it with spit. Then they suck up the half-digested food through the proboscis.

Flies are slightly different. They have a fleshy tip on the end of the proboscis, which they use to sponge up food.

You will need
Velcro	cardboard
glue and scissors	party whistles

1 Cut out some pieces of cardboard. These are the pieces of food you will pick up with your proboscis. Glue a strip of Velcro to each piece of cardboard.

2 Glue another piece of Velcro to the tip of your party whistle.

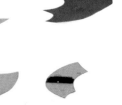

tropical weevil   **giant beetle**   **scalloped oak moth (x 2)**

Weevils have a long snout, or rostrum, with jaws at the end.

3 Scatter the insect food around the floor and try picking it up with the whistle proboscis.

Female mosquitoes feed on blood. They have mouth parts like a syringe, which inject the victim to draw out blood. In some countries, mosquitoes are dangerous because they spread diseases such as malaria.

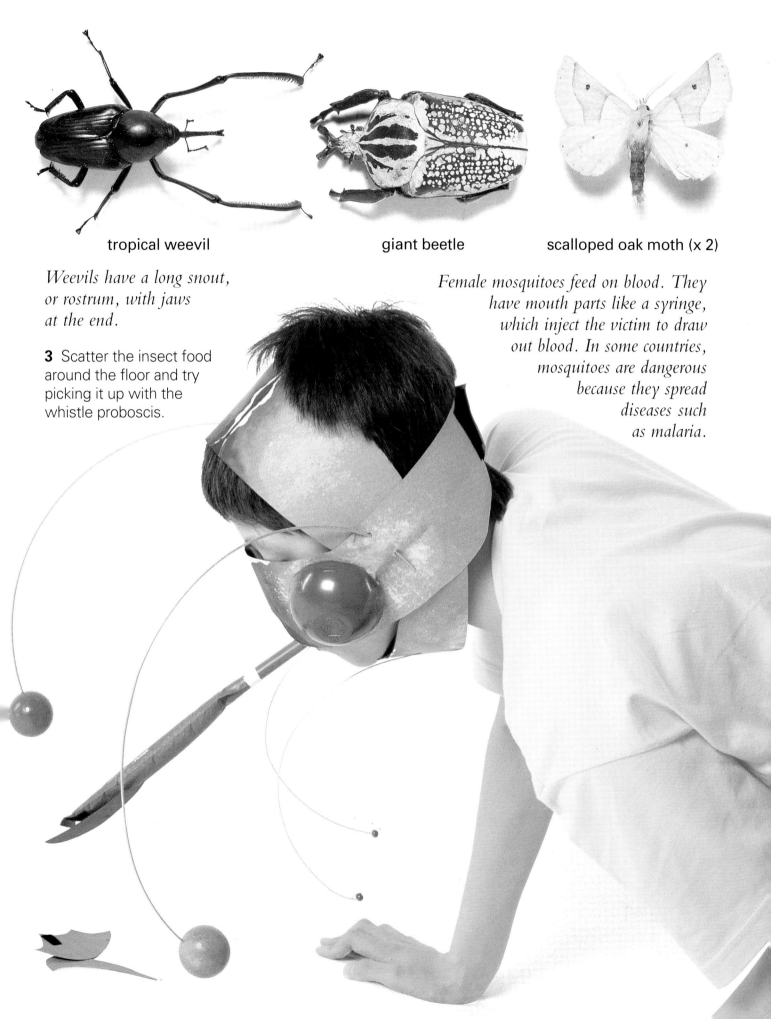

Insects have developed some cunning ways of avoiding their enemies. Stick insects are camouflaged to blend in with their surroundings. Tiny insects are often not spotted by their enemies simply because they are so small.

You will need

two pieces of wire   paint
four paper clips   cardboard
a craft knife or scissors   glue

1 Cut out the cardboard into the shapes shown below. (See the crafty tip on page 13.) Make sure you also punch six holes and cut a slit in the cardboard as indicated.

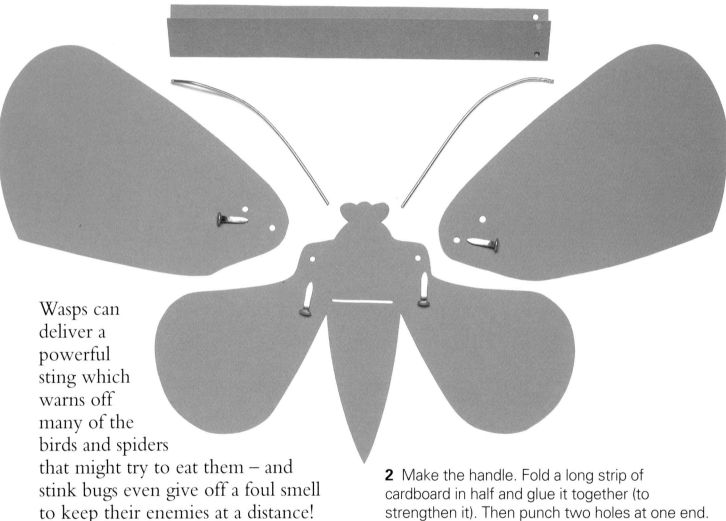

Wasps can deliver a powerful sting which warns off many of the birds and spiders that might try to eat them – and stink bugs even give off a foul smell to keep their enemies at a distance!

2 Make the handle. Fold a long strip of cardboard in half and glue it together (to strengthen it). Then punch two holes at one end.

MAKE it WORK!

Some butterflies and moths have amazing eye patterns on their wings. If they are threatened, they suddenly open up their wings. The enemy is so startled by the size of the huge eyes and the flash of colour that it keeps its distance.

You can make your own model moth, with wings that open to reveal its own secret weapon!

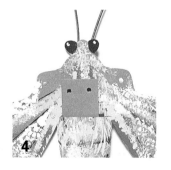

4

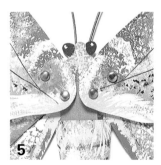

5

3 Paint all four wings in the colours of a moth and then add bright eye patterns to the smaller wings.

Monarch butterflies feed on the sap of the milkweed plant. They take in poisonous chemicals from the milkweed. If a bird eats one of the butterflies, the poisons make it sick. Not surprisingly, birds avoid eating monarch butterflies again!

4 Slide the handle through the slit in the body. Make sure the holes are at the top.

5 Position the large wings on top of the body. Check that the holes line up as shown in the photograph, and fasten them with the paper clips.

6 Now add the wire antennae.

7 When you pull the handle down, the large wings will lift up to show the eye patterns.

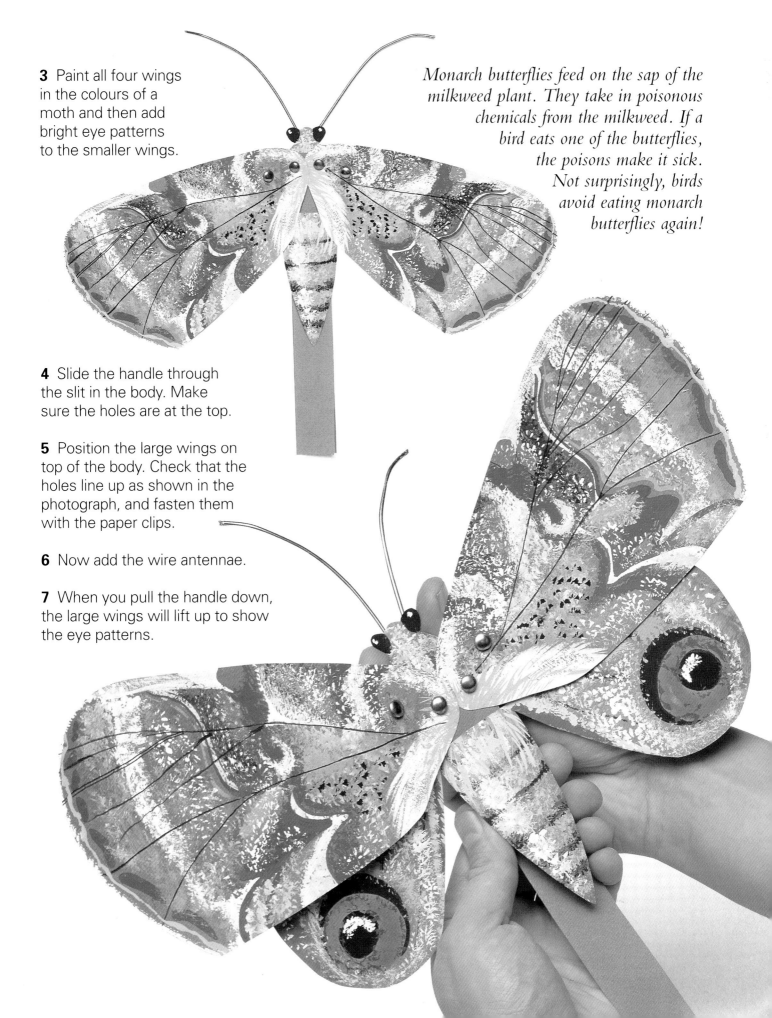

Spiders are one of the insect world's worst enemies. They have two main weapons: good, all-round vision (most spiders have eight eyes!) and poisonous fangs. Many spiders also weave webs to trap insects as they are flying along.

tarantula

Tarantulas are large, hairy spiders with a poisonous bite. The females may live for over 25 years.

MAKE it WORK!

Spiders' webs look fine and delicate, but in fact, they are very strong. They can hold even the biggest, strongest insects. Make your own web and a model spider to go with it. You could add a fly too, based on the model on page 6.

You will need

small nails	glue
fine elastic	cotton
pipe cleaners	polystyrene
round-headed pins	a wooden board

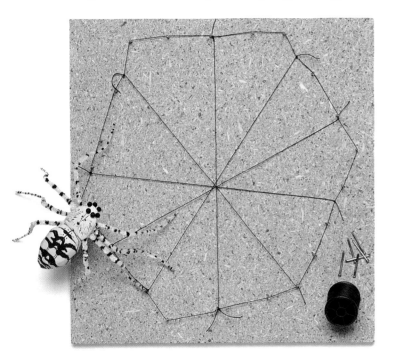

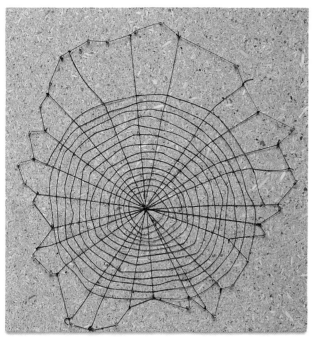

Make the web

1 Hammer the nails into the board in a rough hexagonal (six-sided) shape. Wind a piece of elastic around the outside of the nails.

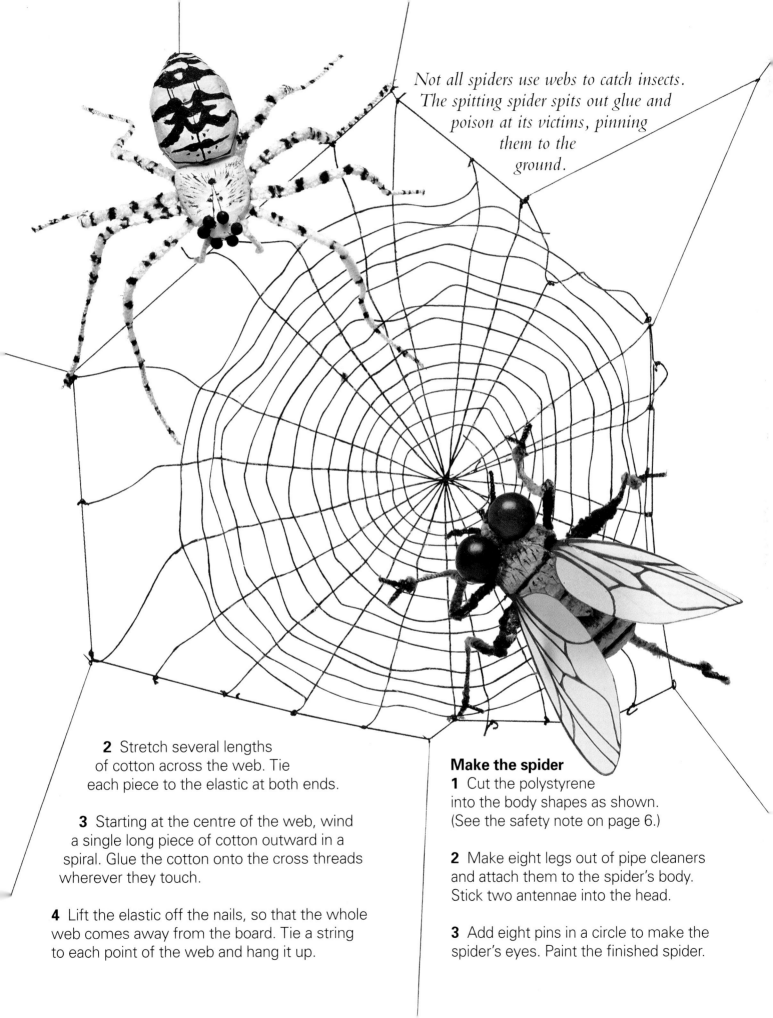

Not all spiders use webs to catch insects. The spitting spider spits out glue and poison at its victims, pinning them to the ground.

2 Stretch several lengths of cotton across the web. Tie each piece to the elastic at both ends.

3 Starting at the centre of the web, wind a single long piece of cotton outward in a spiral. Glue the cotton onto the cross threads wherever they touch.

4 Lift the elastic off the nails, so that the whole web comes away from the board. Tie a string to each point of the web and hang it up.

Make the spider
1 Cut the polystyrene into the body shapes as shown. (See the safety note on page 6.)

2 Make eight legs out of pipe cleaners and attach them to the spider's body. Stick two antennae into the head.

3 Add eight pins in a circle to make the spider's eyes. Paint the finished spider.

Beetles form the largest insect family – over 350,000 species are known. They are well protected, with a pair of hard front wings which fold away to cover the lower thorax and abdomen. They can survive in many different climates, and live everywhere on the earth except for the polar regions.

1 Cut out the cardboard into the shapes shown. (See the tip on page 13 for how to do this best.)

2 Fold the dotted lines on the upper thorax to make an empty box shape. Glue the tabs and add the top to make a complete box. Do the same with the lower thorax and abdomen, covering them with the folded-back wings.

3 Make the head in the same way, but leave it open for the time being.

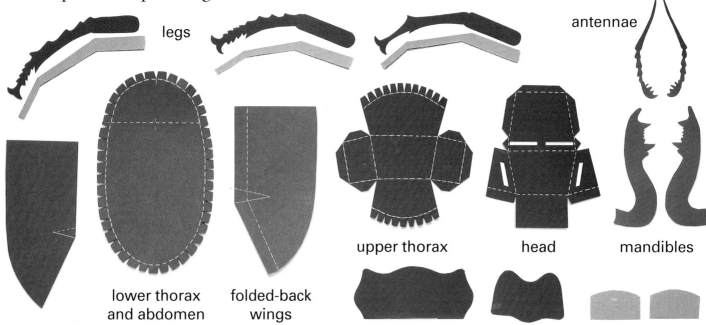

legs

antennae

lower thorax and abdomen

folded-back wings

upper thorax

head

mandibles

MAKE it WORK!
Make a larger-than-life stag beetle, complete with powerful, gripping mandibles, or jaws.

You will need
thumbtacks
coloured cardboard
corrugated cardboard

balsa wood
a craft knife
an elastic band

4 Join the three sections of the model by gluing a piece of corrugated cardboard between them.

5 Glue a piece of balsa wood inside the head. Strengthen the mandibles with corrugated cardboard.

6 Put an elastic band around the mandibles. Slot them through the holes at the side of the head. Fasten them to the balsa wood with thumbtacks.

◀ lower thorax and abdomen

upper thorax   head   mandibles   antennae

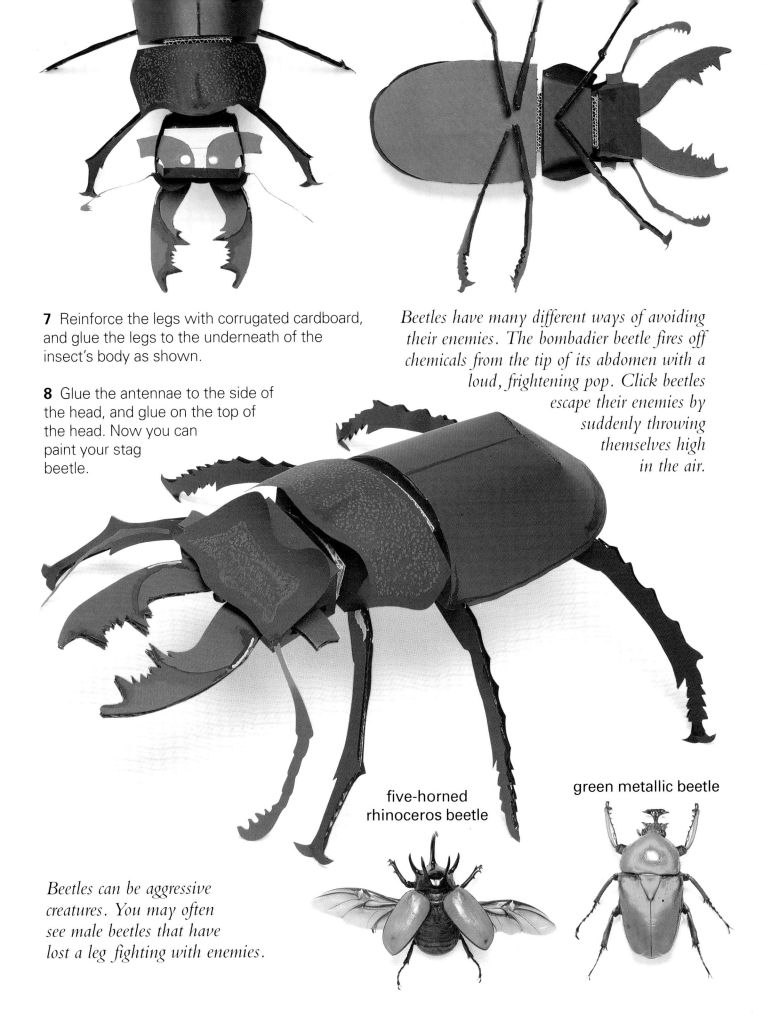

7 Reinforce the legs with corrugated cardboard, and glue the legs to the underneath of the insect's body as shown.

8 Glue the antennae to the side of the head, and glue on the top of the head. Now you can paint your stag beetle.

Beetles have many different ways of avoiding their enemies. The bombadier beetle fires off chemicals from the tip of its abdomen with a loud, frightening pop. Click beetles escape their enemies by suddenly throwing themselves high in the air.

Beetles can be aggressive creatures. You may often see male beetles that have lost a leg fighting with enemies.

five-horned
rhinoceros beetle

green metallic beetle

People often use the word 'bug' as a slang name for all kinds of insects, but in fact, bugs are a special insect group. A true bug has a sharp, jointed feeding tube, rather like a syringe. To eat, the bug spears a piece of food and then sucks it up through the tube. There are about 75,000 different species of bug.

feet

body

legs

antennae

MAKE it WORK!

Pondskaters run across the surface of the water to catch tiny dead or drowning insects. The long, spindly legs are splayed out around the insect's body to spread its weight as far as possible. However, the front legs are a little shorter so that the pondskater can bend down to grab its prey. The legs have a thick covering of water-resistant hairs.

Make your own pondskater and see how it balances on the surface of the water!

You will need
sticky tape
two coloured beads
rubber-solution glue
a polystyrene cutter
  or craft knife
lengths of thin wire
a piece of polystyrene
a thin sheet of balsa wood
acrylic paint or poster paint

Most bugs, such as bedbugs or greenfly, live on land – but some bugs have developed ways of living on, or in, fresh water. Pondskaters can even walk on the surface of water. Water boatmen, on the other hand, live just under the surface. They carry a bubble of air attached to their front and swim along on their back. They have strong back legs which they use like oars to push through the water.

water boatman (x 2)

pondskater (x 2)

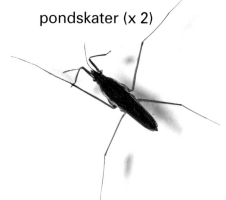

1 Draw the shape of the pondskater's body onto a piece of polystyrene as shown.

2 Cut out the body using a craft knife or a polystyrene cutter. (**Be careful** as you do this! See the safety note on page 6.)

Some insects live in water when they are young, but take to the air once they become adults. Both mayflies and dragonflies start life as water-dwellers called nymphs. They do not grow wings until they reach the end of their life cycles.

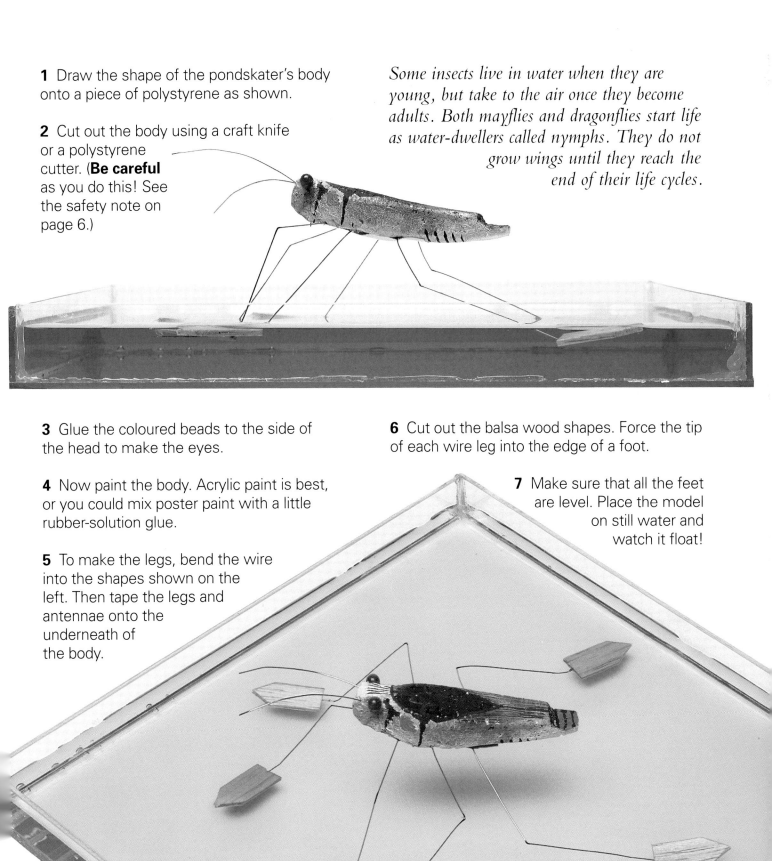

3 Glue the coloured beads to the side of the head to make the eyes.

4 Now paint the body. Acrylic paint is best, or you could mix poster paint with a little rubber-solution glue.

5 To make the legs, bend the wire into the shapes shown on the left. Then tape the legs and antennae onto the underneath of the body.

6 Cut out the balsa wood shapes. Force the tip of each wire leg into the edge of a foot.

7 Make sure that all the feet are level. Place the model on still water and watch it float!

With their colourful wings and graceful flight, butterflies are beautiful creatures. They belong to the same insect group as moths, which has around 150,000 different species. Sadly, some of the rarest species of butterfly are now in danger of **extinction**. They depend upon special habitats to survive, and may die out if the flowers and plants that they need are destroyed by new roads, towns and ways of farming.

You will need

a plastic straw or metal pipe
a craft knife or polystyrene cutter
two tiny beads
a wooden stick
a table-tennis ball
a piece of polystyrene
nuts, bolts and washers

a bradawl
thick cardboard
sandpaper
pipe cleaners
cotton thread

MAKE it WORK!

Make a model butterfly, with wings that move gently up and down. When you come to paint your model, choose one particular kind of butterfly, and try to copy its wing markings.

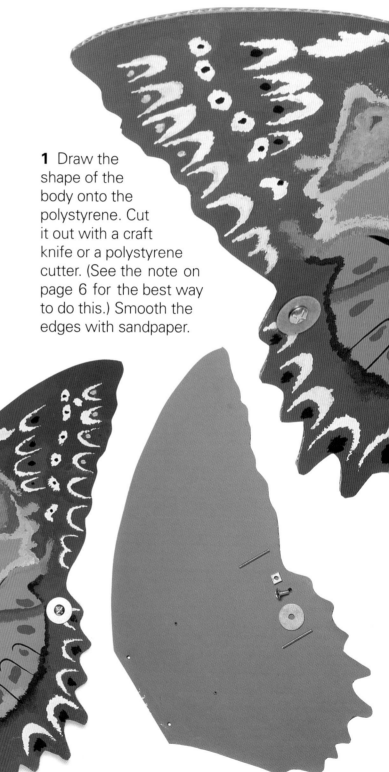

1 Draw the shape of the body onto the polystyrene. Cut it out with a craft knife or a polystyrene cutter. (See the note on page 6 for the best way to do this.) Smooth the edges with sandpaper.

2 Cut the table-tennis ball in half, and glue both halves to the head to make eyes. Add the pipe cleaners to make legs and antennae, and stick a tiny bead onto the end of each antenna.

6 Strengthen the holes through the body with short lengths of metal pipe or a plastic straw.

7 Thread a piece of cotton through the holes on one wing, through the holes in the body and finally through the holes on the second wing. Then tie the two ends of the thread.

8 Make two more holes in the centre of each wing. Thread a piece of cotton through each pair of holes and tie the threads to the stick as shown above.

3 Paint the body with acrylic paints or with poster paints mixed with a little rubber-solution glue.

4 Cut out and paint the two cardboard wings.

5 Make two holes on the inside of each wing. Then use the bradawl to make two holes through the body, lining them up with the wing holes. **Be careful** as you use the bradawl!

9 To make the butterfly move its wings a little, add a balancing weight to the tip of each wing. Attach nuts, bolts and washers as shown, or you could just stick on small lumps of Plasticine.

▼ You can spot each species of butterfly by its wing colouring.

scarce copper

barred giant

spotted forester

swallow tail

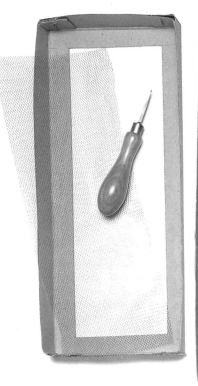

Butterflies and moths don't begin life as graceful flying insects. They start out as caterpillars, and, as they grow up, their bodies go through a complete change, or **metamorphosis**.

The life cycle of a butterfly has four separate stages. First, the adults lay eggs on leaves, and when an egg hatches, a caterpillar comes out. The caterpillar feeds on leaves and grows until it is ready to turn into a **pupa**. The body of the butterfly then develops inside the pupa. At last, a beautiful, adult butterfly emerges. It flies off to look for a mate, and the cycle starts all over again.

You will need

a bradawl
an old shoe box
a craft knife or scissors
two large rubber bands

a camera
fine netting
a twig or dowel

MAKE it WORK!

Watch the amazing process of butterfly and moth metamorphosis for yourself by keeping some caterpillars. Collect them in the spring or autumn, but make sure that you have a good supply of the same leaves as you find them on. Caterpillars have huge appetites! Keep them in a special box, and take photographs of each different stage in the insect's life cycle.

The caterpillar

Caterpillars eat their way out of their eggs, and spend the rest of their short lives munching with their massive jaws. They grow fast, and shed their skins several times to make room for their bodies as they get larger.

The pupa

When a caterpillar is fully grown, it hangs upside down from a branch, loses its final caterpillar skin, and becomes a pupa. The pupa does not eat, but inside its skin all the parts of the adult insect are slowly forming.

Many pupae have a protective covering of thin fibres called a cocoon. We make the material silk from the cocoon of the silk moth.

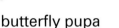

butterfly pupa   giant atlas moth pupa

1 Using the bradawl, punch air-holes all along the sides of the shoe box. Make two larger holes on either side near the top. Poke the twig or dowel through these holes. Some pupae will hang upside down from it.

2 You could also put some soil in the bottom of the box. Some caterpillars prefer to burrow into the soil and become pupae underground.

3 Cut a large window in the lid of the box. Glue or tape a piece of fine netting across the inside of this window.

The adult butterfly
When the butterfly is fully grown, the skin of the pupa splits and the adult comes out into the world. At first, its wings are crumpled and useless, but it spreads them out in the sun until they are dry and hard, and ready for flight.

4 Place your caterpillars and a generous supply of leaves in the bottom of the box. Fasten the lid to the box with the elastic bands.

5 Keep your caterpillar box in a shady place where it is neither too hot nor too cold. Every couple of days, put in some fresh leaves and take out the old ones. Make sure that you clean the box regularly.

We usually see moths flying around after dark. During the day, they prefer to rest on plants and trees.

Many moths are masters of disguise. They often have pale-coloured wings which blend in with the leaves and stems of the plants on which they sit.

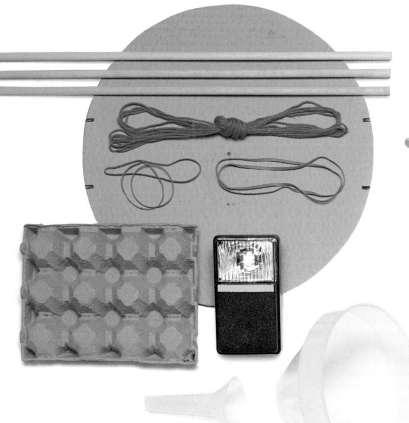

1 First make a tripod by fastening the top of the three canes or dowels together with a rubber band. Sink the base of the tripod into soft earth or grass.

2 Ask an adult to help you cut a hole about 5 cm (2 in.) wide in the middle of the dustbin lid. You will need to use a sharp craft knife – be very careful! If you are using a bucket without a lid, make a cover by cutting a large circle of cardboard with a hole in the middle.

MAKE it WORK!

Although most moths come out at night, they are attracted by light. It's easy to tempt them into a home-made moth trap, where you can look at them closely.

You will need

three long dowels or canes
a plastic garbage can or old bucket
a cardboard egg carton   string
a halogen bicycle lamp   a plastic funnel
a large piece of cardboard   big elastic bands

3 Place the egg carton in the bottom of the garbage can – the moths will settle on it.

4 Cut down the funnel so that it sits neatly in the hole in the garbage can lid. Fasten the lid on the garbage can with large elastic bands. Put the garbage can under the tripod.

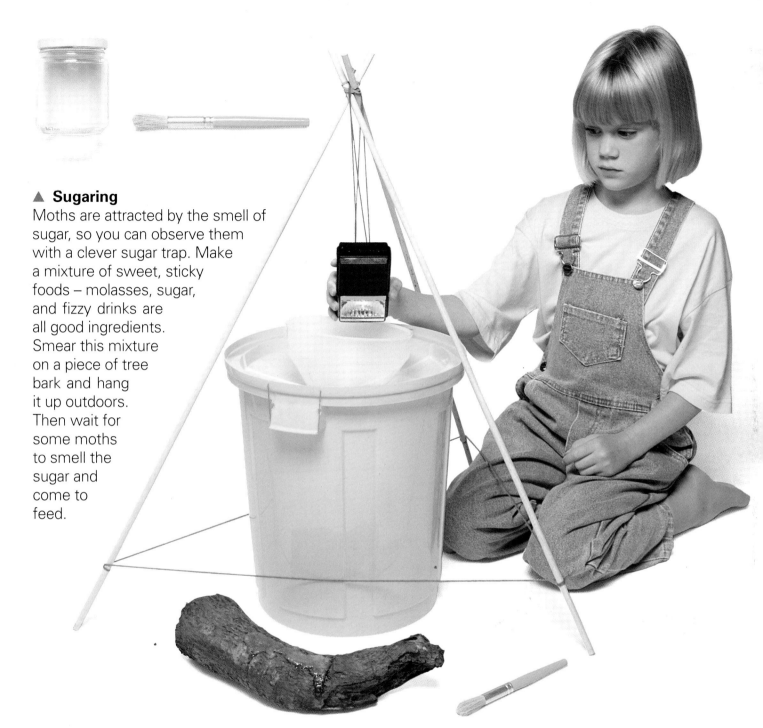

▲ Sugaring

Moths are attracted by the smell of sugar, so you can observe them with a clever sugar trap. Make a mixture of sweet, sticky foods – molasses, sugar, and fizzy drinks are all good ingredients. Smear this mixture on a piece of tree bark and hang it up outdoors. Then wait for some moths to smell the sugar and come to feed.

5 Hang the bicycle lamp on the tripod above the funnel. The moths will be attracted by the light – the brighter the light, the more moths you will collect.

6 Set the moth trap on a warm, summery night. You can look at the moths you have collected the next morning, but be sure to release them afterwards!

Some moths, such as the great oak beauty or the great duss moth, are camouflaged with a mottled brown pattern. Their colouring makes them very difficult to see when they are resting on the bark of trees, and helps them to hide from their enemies.

scalloped oak moth

great duss
moth

36 Flexible Insects

Scientists have recorded over 150,000 species of wasp, bee and ant. They all have one thing in common – the 'wasp waist.' The bottom of the insect's abdomen is separated from the rest of its body by a flexible joint. That way, the insects can move easily around the cramped nests where they live.

You will need
Plasticine
lengths of thin wire
two wooden buttons
a wooden kebab stick

yellow cardboard
tracing paper
tape and cotton
two small beads

1 Cut the cardboard into the shapes shown below. (See our tip on page 13.) Fold along the dotted lines and cut along the solid ones.

2 Glue the tabs to make the three box shapes.

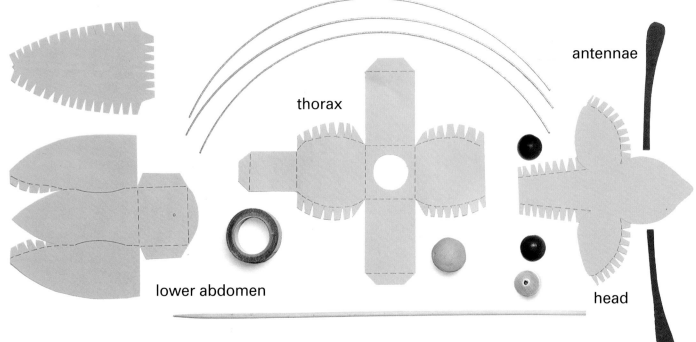

thorax

antennae

lower abdomen

head

MAKE it WORK!
If you've ever been stung by a wasp on a warm summer's day, you might know another reason why wasps have a flexible rear abdomen. The wasp's stinger is in its tail, and the flexible bottom gives it a good aim! You can make your own model wasp, with a realistic wasp waist and moving lower abdomen.

3 Bend two pieces of wire into matching wing shapes. Glue tracing paper to the wire frames, and trim the edges. You can then paint vein patterns on the wings.

4 Poke the kebab stick into the hole at the upper end of the abdomen. Push it through the abdomen and out at the tail to make the stinger.

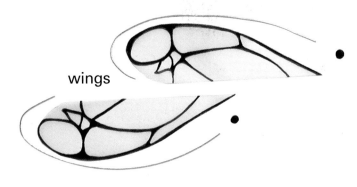

wings

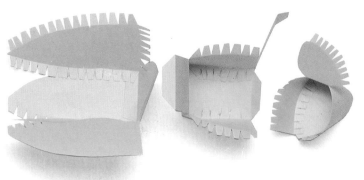

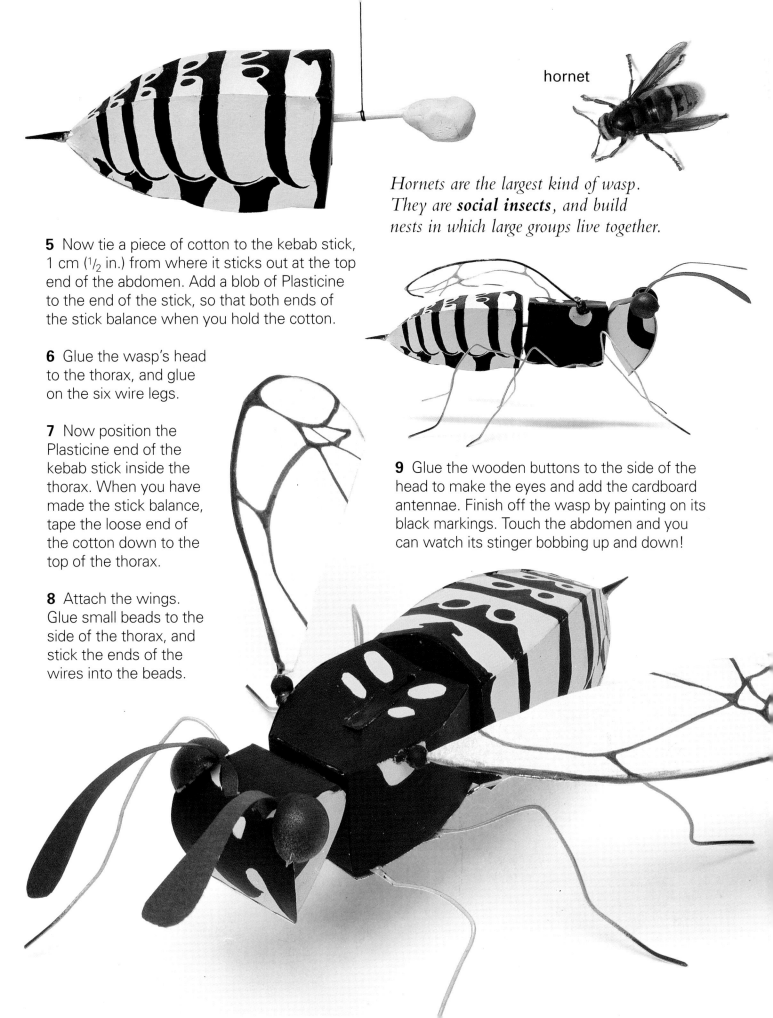

hornet

*Hornets are the largest kind of wasp. They are **social insects**, and build nests in which large groups live together.*

5 Now tie a piece of cotton to the kebab stick, 1 cm (¹/₂ in.) from where it sticks out at the top end of the abdomen. Add a blob of Plasticine to the end of the stick, so that both ends of the stick balance when you hold the cotton.

6 Glue the wasp's head to the thorax, and glue on the six wire legs.

7 Now position the Plasticine end of the kebab stick inside the thorax. When you have made the stick balance, tape the loose end of the cotton down to the top of the thorax.

8 Attach the wings. Glue small beads to the side of the thorax, and stick the ends of the wires into the beads.

9 Glue the wooden buttons to the side of the head to make the eyes and add the cardboard antennae. Finish off the wasp by painting on its black markings. Touch the abdomen and you can watch its stinger bobbing up and down!

Bees live together in large groups. Most of the work is done by female worker bees who gather food, take care of the young and look after the nest, or **hive**.

Worker bees fly from the hive to collect **pollen** and **nectar** from flowers. They take nectar into a special part of the stomach, where they turn it into honey. They store pollen in sacs on their rear legs, and feed it to young bee **grubs**. These bee grubs live inside **honeycombs** – hundreds of six-sided cells made out of wax.

MAKE it WORK!

Make this bee game and turn the page to find out how to play. The players all pretend to be bees, and buzz around trying to collect nectar. The winners are the bees who bring most nectar home to their honeycomb board.

You will need

Plasticine	glue
coloured cardboard	a kebab stick
a pair of compasses	a cocktail stick
a craft knife or scissors	a pencil and ruler

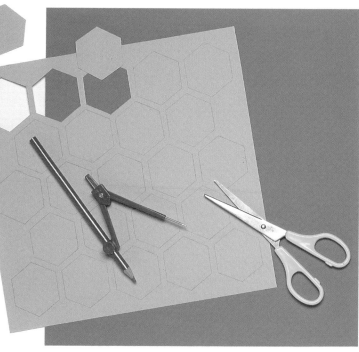

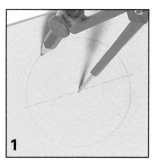
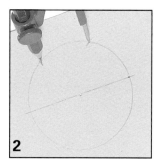

Making a hexagon (six-sided shape) stencil

1 Using the compasses, draw a circle on plain cardboard and rule a line through the middle.

2 Keep the compasses set the same distance apart. Measure out points at this distance all the way around the edge of the circle, as shown above. You will end up with six points.

3 Join the six pencil marks together with a ruler to make a hexagon, and cut it out.

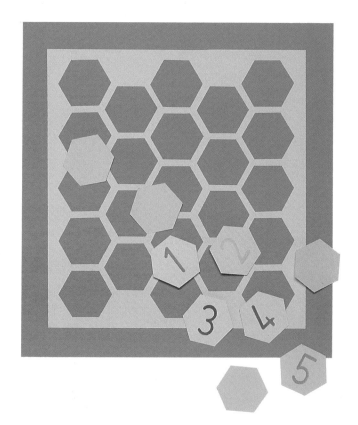

Making the boards and counters

1 Take a large square of cardboard, and, using the hexagon stencil, draw a honeycomb pattern as shown. Cut out all the hexagon shapes.

2 Glue the honeycomb cut-out onto a slightly larger, different-coloured square of cardboard. You will need one of these boards for each team.

3 Take the cut-out hexagons and number them from 1 to 5. Make six of each number. These are the nectar counters you will be collecting.

▼ Making the flowers

1 Using the hexagon shape as the centre, draw and cut out five cardboard flowers.

2 Make cardboard leaf shapes, attached to the hexagon centre. Glue the flowers onto the leaves.

◀ **Making the sun**
Cut out a yellow cardboard sun. Glue it to a kebab stick and add a Plasticine base.

▲ spinner

▼ leaf shape

▶ The nectar counters fit face down inside the flowers.

Making the spinner
Cut out another hexagon. Divide it into six triangles and number them as shown. Push a cocktail stick through the centre.

Playing the bee game

Become a bee yourself and collect as much nectar as you can. At the end of the game, the team with most nectar points wins.

1 Decide where you will play the game. You need a large space – the whole house is ideal, or you can play outside in a park.

2 Divide the players into teams. Each team should have two players, or bees.

3 Ask an adult to place the flowers in different rooms, with the nectar counters face down. Each flower should contain six counters, all with the same number on them. Put the empty hives together in a central room.

4 The first team spins the spinner. One bee then sets out to look for a flower – but it can only take the number of steps shown on the spinner.

The most important bee in a hive is the **queen bee**. *She is a special kind of female bee – larger than all the female workers, and the only female bee who lays eggs. Male bees are called* **drones**. *They do no work and their only job is to mate with the queen in spring and summer. However, in autumn and winter, when there is not much honey to feed on, the females let all the drones starve to death!*

5 The other teams spin in turn. Each bee at the hive spins and shouts out its score until the partner reaches a flower. The collector bee then takes a nectar counter and buzzes straight back to the hive. It puts the counter face down on the hive, without showing it to anyone.

6 Now it's the second bee's turn to collect some nectar. The first bee should set it off in the right direction by doing the waggle dance. (See the opposite page.)

7 The game ends when all the nectar cards have been collected. Each team turns over the cards in its hive and counts up the points. The hive with the highest score wins.

◀ Go straight ahead to find the nectar!

▼ Buzz off to the right to look for a flower!

▶ Fly away at a five-o'clock angle to find high-scoring nectar counters!

Doing the waggle dance

You can tell your partner where to find the nectar by doing the waggle dance, just like a bee. Point yourself in the direction of the nectar, and dance around in a figure-eight, as shown. Do an extra fast waggle when you're directly lined up with nectar!

Waggling tactics

If the first bee found a high-scoring flower, it will want to send the second bee towards that flower. If it found a low scoring flower, it should point the second bee in another direction.

Bees can't talk like humans, but they do have a very effective way of communicating. They tell one another where to go by doing a special dance! They fly in a figure-eight pattern around a central 'waggle' line, which is at an angle to the sun. The waggle line is like a signpost, pointing to the source of the nectar.

Along the waggle line, bees do a special waggling dance. If a bee has found a really good source of nectar, she will do an extra fast, excited dance!

Like bees, ants are social insects, living together in colonies. Different ants have different jobs. The queen lays the eggs. Male ants, or drones, mate with the queen. All the other female ants are the workers: they look after the grubs, build the nest, gather food and keep the nest clean.

MAKE it WORK!

You can make your own ant colony and observe the way ants live at close quarters.

1 First, you should find some ants. The spring months are a good time for collecting them, because they become more active as the weather gets warmer. They usually live hidden under stones or in decaying wood. When you have found a nest, scoop up a few ants, along with some soil, and put them in a collecting jar covered with gauze.

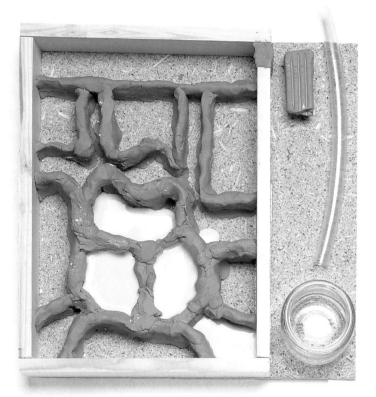

You will need
Plasticine
plaster of Paris
a jam jar with lid
strong wood glue
plastic or rubber tubing
wood for the base and side pieces
a sheet of glass or acrylic and masking tape

2 Ask an adult to help you cut four side pieces and glue them to the board. Make sure the pieces all fit tightly so the ants cannot escape, but make one side piece slightly too short, so that there is a small gap at one corner.

3 Mould the Plasticine to make a pattern of walls and chambers on the chipboard base. Make the walls about 2 cm (1 in.) thick.

4 Mix some plaster of Paris with water in an old jug and pour the liquid into the spaces between the Plasticine walls.

5 Leave the plaster of Paris for 24 hours to set. Then pull out the Plasticine walls, making a network of plaster rooms. Add a thin layer of sand or soil to the base of your box.

6 Put some leaves and soil inside the jar.

7 Ask an adult to help you pierce a small hole in the jam jar lid, and fit the plastic tube through the hole. Seal the hole with Plasticine.

Ants are fairly small insects. The largest ant in the world is from Brazil, but even it measures just 33 mm (1¹/₄ in.) in length. The world's smallest ant comes from Sri Lanka and is a tiny 0.9 mm (¹/₃₂ in.) from head to abdomen.

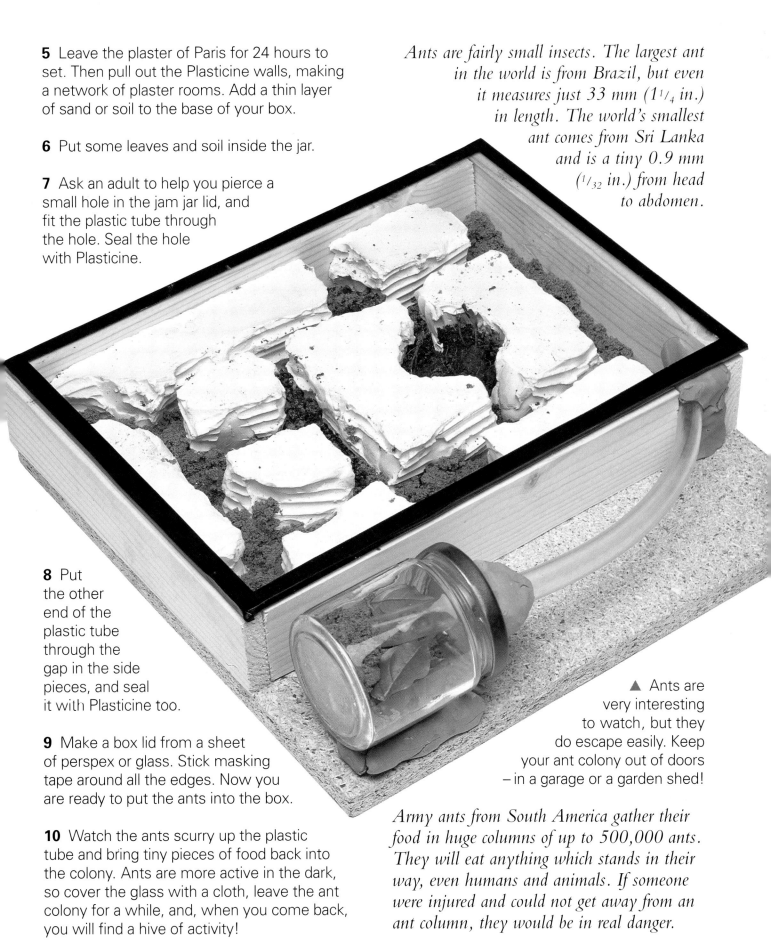

8 Put the other end of the plastic tube through the gap in the side pieces, and seal it with Plasticine too.

9 Make a box lid from a sheet of perspex or glass. Stick masking tape around all the edges. Now you are ready to put the ants into the box.

10 Watch the ants scurry up the plastic tube and bring tiny pieces of food back into the colony. Ants are more active in the dark, so cover the glass with a cloth, leave the ant colony for a while, and, when you come back, you will find a hive of activity!

▲ Ants are very interesting to watch, but they do escape easily. Keep your ant colony out of doors – in a garage or a garden shed!

Army ants from South America gather their food in huge columns of up to 500,000 ants. They will eat anything which stands in their way, even humans and animals. If someone were injured and could not get away from an ant column, they would be in real danger.

Many social insects build special homes where they breed and rear their young. Wasps' nests are started each spring by a single queen, who needs a place to lay her eggs. When the eggs hatch, the adults become workers and help to make the nest larger. When it is finished, it may house up to 500 adults.

MAKE it WORK!

Wasps' nests are made of chewed-up wood fibre. It is mixed with the wasp's saliva to make a kind of pulp. It's hard work for a colony of wasps to make a nest which can be as much as 45 cm (18 in.) long. Build your own wasp's nest, but take a short cut to make the wood pulp!

wasp's nest

You will need
Vaseline
a balloon
white paper
old newspaper
wallpaper paste
a food blender
or hand whisk

1 Blow up the balloon and cover it with a thin layer of Vaseline.

2 Tear up strips of newspaper and stick them onto the balloon using plenty of paste.

3 Tear the white paper into small pieces and soak it in water overnight. Drain and then mash the paper into a pulp with a little more water. Use a hand whisk, or an electric food blender.

Be Careful! Electric food blenders are dangerous, and you should not use one without an adult to help you. **Never** put your hands near the blades.

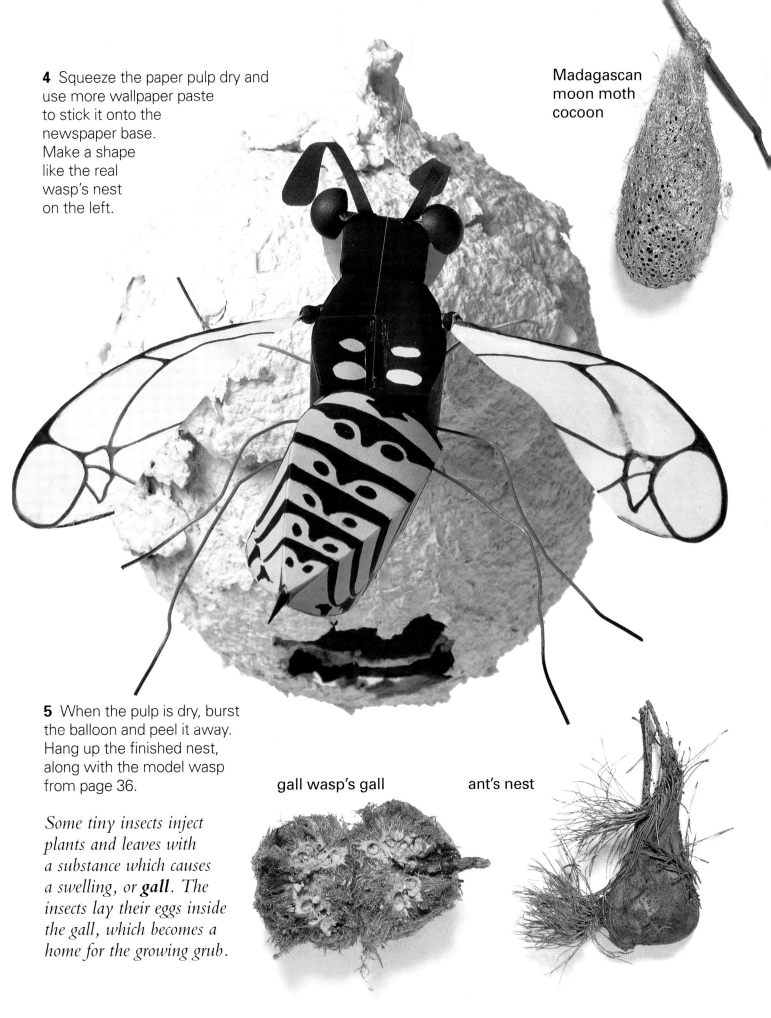

4 Squeeze the paper pulp dry and use more wallpaper paste to stick it onto the newspaper base. Make a shape like the real wasp's nest on the left.

Madagascan moon moth cocoon

5 When the pulp is dry, burst the balloon and peel it away. Hang up the finished nest, along with the model wasp from page 36.

*Some tiny insects inject plants and leaves with a substance which causes a swelling, or **gall**. The insects lay their eggs inside the gall, which becomes a home for the growing grub.*

gall wasp's gall

ant's nest

Abdomen The rear section of an insect's body. It contains the heart, the stomach and the reproductive organs.

Antennae An insect's two feelers, used mainly for smelling and touching. They are attached to the insect's head.

Compound Eye An insect's compound eye has a number of separate lenses, each of which sees at a slightly different angle. (Each human eye, on the other hand, has just one lens.)

Drone A male bee or ant. His only job is to mate with the queen.

Entomologist A scientist who has a special interest in insects.

Exoskeleton The insect's skeleton, which is on the outside of its body (unlike a human skeleton, which is underneath the skin and muscle). The exoskeleton gives the insect extra protection, just like a suit of armour. Some other animals, such as scorpions, also have exoskeletons.

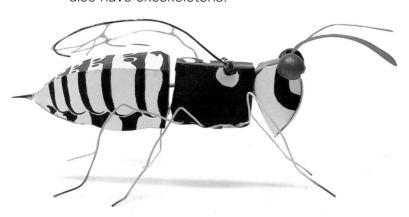

Extinction The complete disappearance of a species of animal or plant from the earth.

Femur The upper part of an insect's leg. It is often the largest section, housing powerful jumping muscles.

Galls Swellings on plants or leaves. They are caused when an insect injects a substance into the plant, which makes part of it swell. Galls provide food and shelter for growing grubs.

Genus The group of insects to which every individual species belongs.

Grubs Young insects, especially beetles, which are still developing into adults. They do not look anything like the adult insect.

Habitat An insect's natural home. An insect's habitat may be under stones, on certain leaves, or inside the bark of trees.

Hive The place where a colony of bees lives and rears its young.

Honeycomb The cells inside a beehive where the honey is stored, and where the young bee grubs live. The honeycomb is made out of the beeswax which the worker bees produce in their abdomens.

Lens A curved section inside an eye which focuses rays of light. The compound eyes of insects may contain many hundreds of separate lenses.

Metamorphosis As some insects grow and develop, their bodies change shape dramatically. These changes are the insect's metamorphosis.

Mandibles The jaws of an insect, used for biting, grinding and crushing food.

Nectar A sweet, sugary liquid produced by flowers. It is collected by bees, butterflies and other insects.

Pheromones Chemicals produced by some insects. They are used to send out messages which other insects detect with their antennae.

Pitfall trap A container, such as a jam jar, which is buried in the ground, with its top at surface level. Insects which scurry along the ground will fall into the trap.

Pollen The yellow dust found inside most flowers. Bees collect pollen, store it in little baskets on their rear legs, and use it to feed their young.

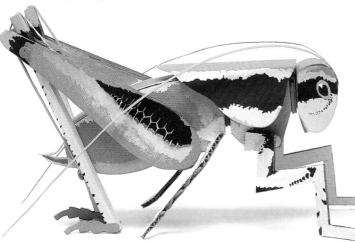

Pooter A simple device that is used by entomologists to collect small insects.

Proboscis A hollow feeding tube which some insects use to suck up liquid foods.

Pupa A central stage in the metamorphosis of an insect. The pupa looks like a small sack, and the adult insect develops inside it.

Queen bee The only bee that can lay eggs. She has a larger abdomen than other bees and lives for much longer. In each hive there is only one queen.

Social insects Insects such as wasps, bees and ants that live together in colonies. They share jobs such as rearing the young and obtaining food.

Species Every different kind of animal is a different species. Over one million species of insects have been recorded by entomologists.

Thorax The middle section of an insect's body. The insect's legs and wings are joined to the thorax.

abdomens 6, 7, 27, 36
antennae 6, 12-13
ants 5, 42-43

bedbugs 28
bees 38-41
  communication of 41
beetles 4, 5, 7, 20, 21,
  26-27
  bombadier beetle 27
  click beetle 27
  giant beetle 21
  green metallic beetle 27
  khao yai beetle 7
  rhinoceros beetle 5, 27
  Sumatran beetle 5
breeding 4, 32, 40, 44, 45
bug boxes 4, 9
butterflies 4, 22, 23, 30-31
butterfly nets 4, 9

camouflage 22, 33, 34, 35
caterpillars 32, 33
centipedes 7
cocoons 33, 45
crickets 14, 15

diseases, spread by insects
  20, 21
dragonflies 18-19, 20, 29
drones 40, 42

earwigs 5
enemies of insects 22-23,
  24, 25, 27

entomologists 4
exoskeleton 17
eyes 6, 10-11, 18, 24

feeding 5, 8, 20-21, 23, 28,
  32, 35, 38, 42, 43
fleas 14, 20
flies 4, 16, 17, 19, 20

galls 45
grasshoppers 14, 15, 20
greenflies 28
grubs 38, 42, 45

habitats 4, 5, 26, 30
hives 38, 40
honey 38, 40
honeycombs 38, 39
hornets 37
horse-flies 11, 17

insect spotter's code 8

jaws 20, 21, 26, 32

malaria 21
mandibles see jaws
mayflies 29
metamorphosis 32-33
millipedes 7
monarch butterflies 23
mosquitoes 19, 21
moth traps 34

moths 13, 20, 21, 22, 23, 30,
  32, 34-35
muscles 14, 16, 17

nectar 38, 39, 41
nests 36, 37, 38, 42, 44, 45

pheromones 13
pitfall traps 8
pollen 38
pondskaters 28, 29
pooters 4, 8, 9
proboscises 20
pupae 32, 33

silk moth 33
spiders 7, 22, 24-25
stick insects 22
stink bugs 22

tarantulas 24
thorax 4, 16

veins 16

wasps 22, 36-37, 44
water boatmen 28, 29
water bugs 28-29
weevils 21
wings 16, 17, 18, 26, 29, 30,
  31, 33, 34